Editor
Lorin E. Klistoff, M.A.

Managing Editor
Karen Goldfluss, M.S. Ed.

Editor-in-Chief
Sharon Coan, M.S. Ed.

Cover Artist
Barb Lorseyedi

Art Manager
Kevin Barnes

Art Director
CJae Froshay

Imaging
James Edward Grace
Rosa C. See

Product Manager
Phil Garcia

Publishers
Rachelle Cracchiolo, M.S. Ed.
Mary Dupuy Smith, M.S. Ed.

Author
Mary Rosenberg

Teacher Created Materials, Inc.
6421 Industry Way
Westminster, CA 92683
www.teachercreated.com
ISBN-0-7439-8603-2

©2004 *Teacher Created Materials, Inc.*
Made in U.S.A.

The classroom teacher may reproduce copies of materials in this book for classroom use only. The reproduction of any part for an entire school or school system is strictly prohibited. No part of this publication may be transmitted, stored, or recorded in any form without written permission from the publisher.

Table of Contents

Introduction 3
Practice 1:   Place Value Blocks (Hundreds) 4
Practice 2:   More Place Value Blocks (Hundreds)... 5
Practice 3:   Place Value (Thousands).. 6
Practice 4:   Expanded Notation (Thousands) 7
Practice 5:   Place Value (Thousands).. 8
Practice 6:   Less and More.. 9
Practice 7:   Place Value Blocks (Thousands).. 10
Practice 8:   Place Value Blocks (Thousands).. 11
Practice 9:   Comparing Numbers... 12
Practice 10: Adding Four Digits Without Regrouping 13
Practice 11: Subtracting Four Digits Without Regrouping 14
Practice 12: Adding and Subtracting Four Digits Without Regrouping 15
Practice 13: Adding Four Digits with Regrouping 16
Practice 14: Subtracting Four Digits with Regrouping 17
Practice 15: Place Value Riddles (Thousands) 18
Practice 16: Rounding Numbers 19
Practice 17: Rounding Numbers 20
Practice 18: Rounding More Numbers 21
Practice 19: Expanded Notation (Ten Thousands) 22
Practice 20: Place Value (Ten Thousands) 23
Practice 21: More Practice with Less and More... 24
Practice 22: Adding Five Digits Without Regrouping 25
Practice 23: Subtracting Five Digits Without Regrouping 26
Practice 24: Adding and Subtracting Five Digits Without Regrouping 27
Practice 25: Adding Five Digits with Regrouping... 28
Practice 26: Column Addition 29
Practice 27: Subtracting Five Digits with Regrouping and Comparing Numbers ... 30
Practice 28: Using Expanded Form with Larger Numbers 31
Practice 29: Using Expanded Form with Words... 32
Practice 30: Using Decimals.. 33
Practice 31: Using a Decimal Point.. 34
Practice 32: Comparing Decimals and Adding and Subtracting Decimals 35
Practice 33: Adding and Subtracting Decimals (Money) 36
Practice 34: Money Riddles 37
Practice 35: Working with Integers 38
Practice 36: Adding and Subtracting with Integers 39
Test Practice Pages.. 40
Answer Sheet 45
Answer Key... 46

Introduction

The old adage "practice makes perfect" can really hold true for your child and his or her education. The more practice and exposure your child has with concepts being taught in school, the more success he or she is likely to find. For many parents, knowing how to help your children can be frustrating because the resources may not be readily available. As a parent, it is also difficult to know where to focus your efforts so that the extra practice your child receives at home supports what he or she is learning in school.

This book has been designed to help parents and teachers reinforce basic skills with children in grade 3. *Practice Makes Perfect: Place Value (Grade 3)* reviews place value. While it would be impossible to include all concepts taught in grade 3 in this book, the following basic objectives are reinforced through practice exercises. These objectives support math standards established on a district, state, or national level. (Refer to Table of Contents for specific objectives of each practice page.)

- place value to ten thousands
- identifying the value for a given digit
- expanded notation using numbers, words, and a combination of numbers and words
- comparing numbers
- adding four and five digits
- subtracting four and five digits
- column addition
- solving word problems
- rounding to the nearest ten, hundred, and thousand
- decimals to the tenths and hundredths
- adding and subtracting money
- integers

There are 36 practice pages. (*Note:* Have children show all work where computation is necessary to solve a problem. For multiple-choice responses on practice pages, children can circle the answer.) Following the practice pages are five test practices. These provide children with multiple-choice test items to help prepare them for standardized tests administered in schools. To correct the test pages and the practice pages in this book, use the answer key provided on pages 46, 47, and 48.

How to Make the Most of This Book

Here are some useful ideas for optimizing the practice pages in this book:

- Set aside a specific place in your home to work on the practice pages. Keep it neat and tidy with materials on hand.

- Set up a certain time of day to work on the practice pages. This will establish consistency. Look for times in your day or week that are less hectic and more conducive to practicing skills.

- Keep all practice sessions with your child positive and more constructive. If your child is having difficulty understanding what to do or how to get started, work through the first problem with him or her.

- Review the work your child has done. This serves as reinforcement and provides further practice.

- Pay attention to the areas in which your child has the most difficulty. Provide extra guidance and exercises in those areas. Allowing children to use drawings and manipulatives, such as coins, tiles, game markers, or flash cards, can help them grasp difficult concepts more easily.

Place Value Blocks (Hundreds)

Practice 1

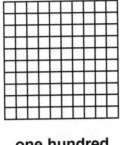
one hundred
100

ten
10

one
1

Directions: Write the number.

Example

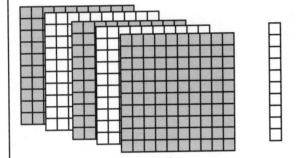

__5__ hundreds + __1__ ten + __0__ ones

__500__ + __10__ + __0__

__510__

1.

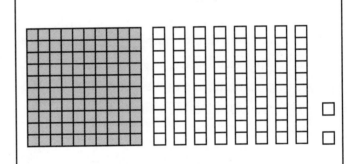

____ hundred + ____ tens + ____ ones

____ + ____ + ____

_____

2.

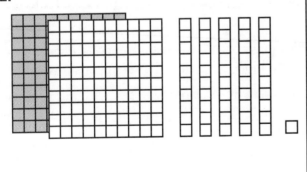

____ hundreds + ____ tens + ____ one

____ + ____ + ____

_____

3.

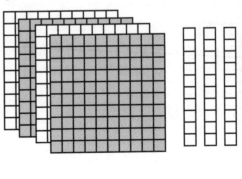

____ hundreds + ____ tens + ____ ones

____ + ____ + ____

_____

More Place Value Blocks (Hundreds)

Practice 2

Directions: Write the number.

1.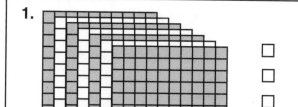

   _____ hundreds + _____ tens + _____ ones

   _____ + _____ + _____

   _____

2.

   _____ hundreds + _____ tens + _____ ones

   _____ + _____ + _____

   _____

3.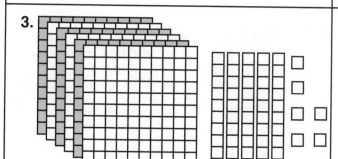

   _____ hundreds + _____ tens + _____ ones

   _____ + _____ + _____

   _____

4.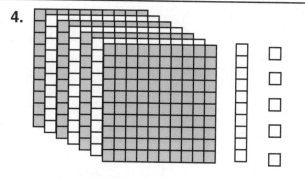

   _____ hundreds + _____ ten + _____ ones

   _____ + _____ + _____

   _____

5.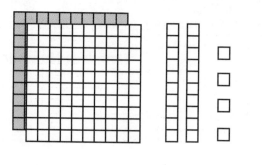

   _____ hundreds + _____ tens + _____ ones

   _____ + _____ + _____

   _____

6.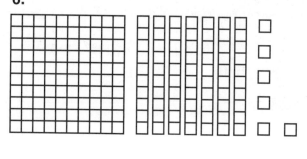

   _____ hundred + _____ tens + _____ ones

   _____ + _____ + _____

   _____

Place Value (Thousands)

Practice 3

> In the number 5,942, the 5 has the greatest value. It is in the *thousands* place.

Thousands	Hundreds	Tens	Ones
1,	0	6	7
5,	9	4	2

> In the number 1,067, 7 is the largest digit. It is in the *ones* place.

Directions: Write the value of the circled number.

Example: 3,61⑦ _____ones_____

1. ⑦,710 _____
2. 5,①88 _____
3. 9,8⑦7 _____
4. ④,820 _____
5. 8,2⑤3 _____

6. 2,⑥91 _____
7. 3,26⑨ _____
8. ①,498 _____
9. ⑥,260 _____
10. 3,①34 _____

Directions: Circle the largest digit. Write its value.

11. 1,480 _____
12. 1,401 _____
13. 7,653 _____
14. 6,581 _____
15. 3,481 _____

16. 9,326 _____
17. 7,363 _____
18. 5,984 _____
19. 8,601 _____
20. 7,181 _____

Directions: Using each set of numbers, make both the largest number possible and the smallest number possible. (*Note:* Zero can not be used in the thousands place.)

	Largest Number	**Smallest Number**
21. (2, 1, 5, 0)	_____	_____
22. (3, 3, 4, 2)	_____	_____
23. (3, 7, 0, 6)	_____	_____
24. (6, 5, 3, 6)	_____	_____
25. (5, 7, 2, 2)	_____	_____
26. (6, 4, 4, 8)	_____	_____
27. (5, 1, 2, 4)	_____	_____
28. (6, 8, 7, 9)	_____	_____

Expanded Notation (Thousands)

Practice 4

Directions: Rewrite each number in expanded notation using only numbers.
Example: 6,496   6,000 + 400 + 90 + 6

1. 5,523 _____
2. 6,124 _____
3. 7,439 _____
4. 6,240 _____
5. 8,205 _____

Directions: Rewrite each number in expanded notation using numbers and words.
Example: 8,547   8 thousands + 5 hundreds + 4 tens + 7 ones

6. 3,411 _____
7. 2,357 _____
8. 7,163 _____
9. 1,963 _____
10. 4,534 _____

Directions: Rewrite each number in expanded notation using only words.
Example: 8,098   eight thousands + nine tens + eight ones

11. 1,507 _____
12. 5,909 _____
13. 2,618 _____
14. 7,087 _____
15. 4,927 _____

Directions: Rewrite each number word in standard form.
Example: two thousand, seven hundred forty-seven   2,747

16. eight thousand, five hundred nineteen _____
17. five thousand, eight hundred thirty-two _____
18. one thousand, five hundred sixty-six _____
19. two thousand, six hundred ninety-three _____
20. nine thousand, two hundred forty-six _____

Place Value (Thousands)

Practice 5

When writing numbers or number words, use a comma to separate the thousands from the hundreds.

	Thousands	Hundreds	Tens	Ones
7,000 + 700 + 30 + 6	7 ,	7	3	6
8 hundreds + 1 ten + 0 ones		8	1	0
41			4	1
five thousand, six hundred twelve	5 ,	6	1	2

Directions: Write each number in standard form. Circle the numbers where a comma was used to separate the thousands from the hundreds.

Example: seven thousand, one hundred sixty-three   (7,1)63

1. 9981   _____
2. 500 + 80 + 8   _____
3. 7 thousands + 5 hundreds + 3 tens + 9 ones   _____
4. four hundred thirty-eight   _____
5. 6,000 + 200 + 10 + 8   _____
6. 1011   _____
7. 5 thousands + 7 hundreds + 4 tens + 2 ones   _____
8. 8 hundreds + 4 tens + 2 ones   _____
9. two thousand, five hundred ninety-five   _____

Directions: Circle the largest number. Then draw a line under the smallest number.

Example: (5,910)	591	59	5	14.	107	1,077	10	1
10.	616	61	6,168	6	15.	4,231	42	423	4
11.	4	47	474	4,742	16.	675	67	6,751	6
12.	261	2	2,613	26	17.	751	7,518	7	75
13.	4,393	43	423	4	18.	7,848	7	784	78

19. Write a number in the thousands with a 7 in the tens place.   _____

20. Write a number in the thousands with a 4 in the thousands place.   _____

Practice 6

Directions: Write the numbers that are one less and one more.

1. _____ 785 _____
2. _____ 752 _____
3. _____ 303 _____
4. _____ 326 _____
5. _____ 453 _____
6. _____ 577 _____
7. _____ 218 _____
8. _____ 187 _____
9. _____ 495 _____
10. _____ 943 _____
11. _____ 353 _____
12. _____ 509 _____
13. _____ 221 _____
14. _____ 785 _____
15. _____ 468 _____
16. _____ 879 _____
17. _____ 124 _____
18. _____ 629 _____

Directions: Write the numbers that are ten less and ten more.

19. _____ 873 _____
20. _____ 797 _____
21. _____ 681 _____
22. _____ 549 _____
23. _____ 501 _____
24. _____ 174 _____
25. _____ 829 _____
26. _____ 451 _____
27. _____ 340 _____
28. _____ 804 _____
29. _____ 509 _____
30. _____ 555 _____
31. _____ 125 _____
32. _____ 982 _____
33. _____ 740 _____
34. _____ 637 _____
35. _____ 330 _____
36. _____ 284 _____

Directions: Write the numbers that are ten less and ten more.

37. _____ 5,096 _____
38. _____ 5,955 _____
39. _____ 3,794 _____
40. _____ 9,930 _____
41. _____ 4,812 _____
42. _____ 8,162 _____
43. _____ 1,867 _____
44. _____ 3,269 _____
45. _____ 9,542 _____
46. _____ 5,110 _____
47. _____ 1,259 _____
48. _____ 7,409 _____
49. _____ 8,304 _____
50. _____ 4,936 _____
51. _____ 3,766 _____
52. _____ 2,008 _____
53. _____ 1,905 _____
54. _____ 1,262 _____

Directions: Write the numbers that are one hundred less and one hundred more.

55. _____ 5,132 _____
56. _____ 7,452 _____
57. _____ 3,003 _____
58. _____ 3,264 _____
59. _____ 5,357 _____
60. _____ 7,218 _____
61. _____ 1,874 _____
62. _____ 9,594 _____
63. _____ 3,353 _____
64. _____ 5,092 _____
65. _____ 2,496 _____
66. _____ 6,113 _____
67. _____ 7,216 _____
68. _____ 2,060 _____
69. _____ 5,746 _____
70. _____ 6,829 _____
71. _____ 9,888 _____
72. _____ 7,819 _____

Place Value Blocks (Thousands)

Practice 7

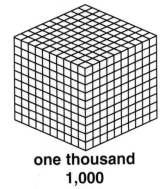
one thousand
1,000

one hundred
100

ten
10

one
1

Directions: Write the number.

Example

__1__ thousand __1__ hundred __2__ tens __1__ one

__1,000__ + __100__ + __20__ + __1__

__1,121__

1.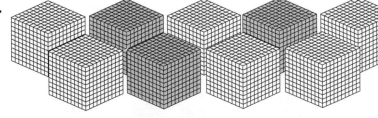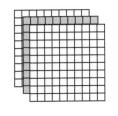

_____ thousands _____ hundreds _____ tens _____ ones

_____ + _____ + _____ + _____

_____

2.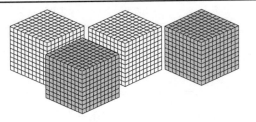

_____ thousands _____ hundreds _____ tens _____ ones

_____ + _____ + _____ + _____

_____

Place Value Blocks (Thousands)

Practice 8

Directions: Write the number.

1.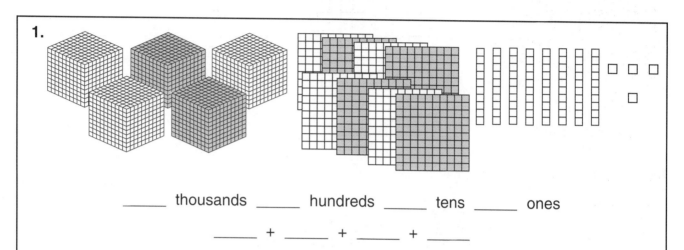

   _____ thousands _____ hundreds _____ tens _____ ones

   _____ + _____ + _____ + _____

   _____

2.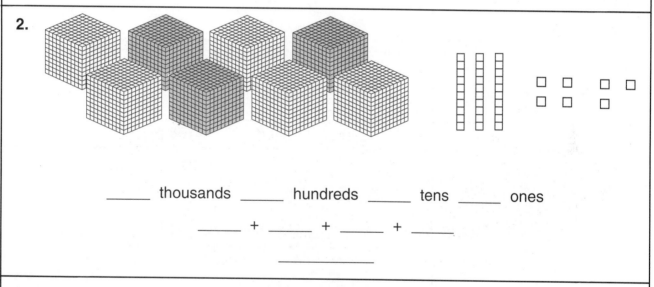

   _____ thousands _____ hundreds _____ tens _____ ones

   _____ + _____ + _____ + _____

   _____

3.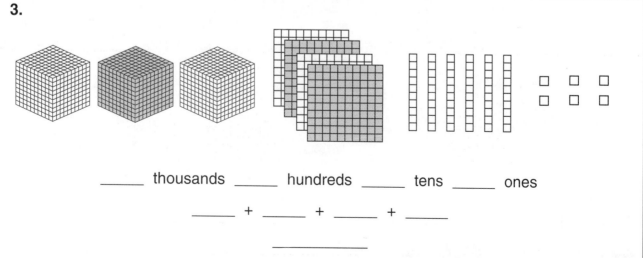

   _____ thousands _____ hundreds _____ tens _____ ones

   _____ + _____ + _____ + _____

   _____

Comparing Numbers

Practice 9

Directions: Use the symbols > (greater than) and < (less than) to compare each set of numbers. Complete each sentence.

Example 9,879 > 9,047 9,879 is <u>greater</u> than 9,047.	1. 8,068 ◯ 8,714 8,068 is _____ than 8,714.	2. 9,748 ◯ 9,401 9,748 is _____ than 9,401.
3. 6,712 ◯ 6,233 6,712 is _____ than 6,233.	4. 5,324 ◯ 5,491 5,324 is _____ than 5,491.	5. 8,551 ◯ 8,753 8,551 is _____ than 8,753.
6. 2,585 ◯ 2,306 2,585 is _____ than 2,306.	7. 1,132 ◯ 1,906 1,132 is _____ than 1,906.	8. 9,149 ◯ 9,925 9,149 is _____ than 9,925.
9. 7,385 ◯ 7,636 7,385 is _____ than 7,636.	10. 8,102 ◯ 8,530 8,102 is _____ than 8,530.	11. 5,758 ◯ 5,734 5,758 is _____ than 5,734.
12. 1,866 ◯ 1,432 1,866 is _____ than 1,432.	13. 5,209 ◯ 5,953 5,209 is _____ than 5,953.	14. 6,224 ◯ 6,249 6,224 is _____ than 6,249.
15. 6,508 ◯ 6,714 6,508 is _____ than 6,714.	16. 1,737 ◯ 1,192 1,737 is _____ than 1,192.	17. 1,894 ◯ 2,712 1,894 is _____ than 2,712.

Directions: Complete each number pattern.

18. 49, 59, 69, _____, _____, _____, _____, _____, _____, _____

19. 12, 22, 32, _____, _____, _____, _____, _____, _____, _____

20. 409, 419, 429, _____, _____, _____, _____, _____, _____, _____

21. 975, 985, 995, _____, _____, _____, _____, _____, _____, _____

22. 3,857   3,867   3,877   _____, _____, _____, _____, _____, _____

Adding Four Digits Without Regrouping

Practice 10

Directions: Solve each problem.

1.	2,559 + 2,130	2.	6,325 + 3,660	3.	3,449 + 5,540	4.	3,130 + 4,028
5.	4,078 + 1,910	6.	2,664 + 4,120	7.	3,423 + 2,433	8.	6,661 + 3,230
9.	1,921 + 3,065	10.	5,594 + 4,103	11.	5,393 + 4,600	12.	3,818 + 3,051
13.	1,922 + 7,077	14.	5,662 + 2,007	15.	2,878 + 1,001	16.	1,979 + 7,020
17.	4,896 + 1,103	18.	3,532 + 5,331	19.	1,000 + 8,759	20.	6,595 + 2,103

Directions: Write each answer in the correct category.

Answer > 5,000	Answer < 5,000

© Teacher Created Materials, Inc. #8603 Practice Makes Perfect: Place Value 13

Subtracting Four Digits Without Regrouping

Practice 11

Directions: Solve each problem.

1. 7,071
   − 4,021

2. 9,634
   − 1,031

3. 7,769
   − 2,402

4. 6,944
   − 5,502

5. 8,455
   − 6,000

6. 7,977
   − 5,715

7. 7,857
   − 2,327

8. 9,215
   − 1,200

9. 8,663
   − 7,243

10. 9,183
    − 6,062

11. 8,738
    − 4,735

12. 7,461
    − 5,410

13. 9,380
    − 4,340

14. 8,609
    − 5,604

15. 9,248
    − 5,218

16. 8,315
    − 6,310

17. 8,816
    − 8,813

18. 6,969
    − 3,059

19. 7,742
    − 6,541

20. 9,174
    − 5,162

21. 9,228
    − 3,228

22. 6,287
    − 4,277

23. 8,704
    − 6,604

24. 8,183
    − 5,030

Directions: Write the answers from each group of problems in order from smallest to largest. Then circle the odd answers.

Group 1 (Problems 1–5): _____, _____, _____, _____, _____

Group 2 (Problems 6–10): _____, _____, _____, _____, _____

Group 3 (Problems 11–15): _____, _____, _____, _____, _____

Group 4 (Problems 16–20): _____, _____, _____, _____, _____

Group 5 (Problems 21–24): _____, _____, _____, _____

Practice 12

Directions: Solve each problem.

1. 6,411 + 2,583
2. 2,958 − 1,903
3. 8,913 − 8,503
4. 8,777 − 7,760

5. 3,249 + 6,110
6. 7,096 − 5,086
7. 7,624 − 7,504
8. 9,575 − 9,070

9. 4,817 − 1,305
10. 3,426 + 2,460
11. 2,355 + 3,524
12. 4,968 − 1,902

13. 3,021 + 4,077
14. 9,167 − 5,163
15. 8,197 − 4,022
16. 9,804 − 8,503

17. 2,464 + 1,135
18. 1,038 + 3,531
19. 8,776 − 8,225
20. 3,779 − 3,549

Directions: Use each set of numbers to make an addition problem or a subtraction problem.

Example: 8, 3, 6, 1 8,361 + 1,638 9,999	**21.** 2, 9, 0, 7	**22.** 6, 7, 4, 9
23. 1, 4, 6, 0	**24.** 3, 8, 7, 9	**25.** 8, 7, 4, 5

Adding Four Digits with Regrouping

Practice 13

Directions: Solve each problem.

1. **A** 3,498 + 3,310	2. **B** 2,576 + 4,890	3. **E** 1,684 + 5,931	4. **H** 3,741 + 2,496
5. **I** 8,657 + 1,492	6. **J** 5,298 + 1,055	7. **M** 5,489 + 2,685	8. **N** 8,590 + 1,018
9. **O** 1,375 + 8,183	10. **P** 1,917 + 7,923	11. **Q** 8,166 + 1,184	12. **R** 4,601 + 2,475
13. **S** 8,228 + 1,162	14. **T** 7,986 + 1,463	15. **U** 1,273 + 7,546	16. **V** 5,953 + 1,267
17. **W** 1,745 + 5,614	18. **X** 1,940 + 2,842	19. **Y** 2,365 + 4,727	20. **Z** 1,218 + 6,375

Directions: Write the letter that goes with each sum on the line.

___ ___ ___ ___ ___ ___ ___
9,608 8,819 8,174 7,466 7,615 7,076 9,390

___ ___ ___
6,808 7,076 7,615

___ ___ ___ ___ ___ ___ ___ ___ ___ ___!
7,615 7,220 7,615 7,076 7,092 7,359 6,237 7,615 7,076 7,615

Practice 14

Directions: Subtract.

1. 6,360 − 5,324
2. 4,110 − 1,090
3. 7,283 − 3,799
4. 1,690 − 1,428
5. 3,122 − 2,514
6. 8,814 − 6,605
7. 9,819 − 8,331
8. 8,895 − 7,746
9. 6,159 − 3,654
10. 7,276 − 3,247
11. 4,190 − 3,615
12. 2,223 − 2,105
13. 4,660 − 1,481
14. 5,898 − 1,409
15. 7,240 − 2,343
16. 9,487 − 5,881
17. 9,600 − 4,872
18. 6,972 − 5,979
19. 3,379 − 1,728
20. 9,169 − 1,428

Directions: How long did each inventor live? Write and solve each problem.

21. Alexander Graham Bell
    Born: 1874
    Died: 1922
    Age: _____

22. George Washington Carver
    Born: 1864
    Died: 1943
    Age: _____

23. Thomas Alva Edison
    Born: 1847
    Died: 1931
    Age: _____

24. Samuel F. B. Morse
    Born: 1791
    Died: 1872
    Age: _____

25. Henry Ford
    Born: 1863
    Died: 1947
    Age: _____

26. Wilbur Wright
    Born: 1867
    Died: 1912
    Age: _____

Place Value Riddles (Thousands)

Practice 15

Directions: Read and solve the number for each riddle.

1. 0, 1, 1, 2 The digit with the largest value is in the tens place. The two odd numbered digits are next to each other. The 1 is first. What is the number? _____	**2.** 4, 5, 9, 9 The two 9's are next to each other. The 4 has the greatest value. The 5 has the least value. What is the number? _____
3. 1, 2, 6, 7 The 7 has the greatest value. The 1 has the least value. The 2 is not in the hundreds place. What is the number? _____	**4.** 0, 2, 7, 8 The 2 has the greatest value. The 8 is in the tens place. The 7 is worth more than the 8 but less than the 2. What is the number? _____
5. 3, 4, 5, 7 The 7 is worth more than the 4. The 3 is worth more than the 7. The 5 is worth more than the 3. What is the number? _____	**6.** 1, 3, 4, 5 The 1 is in the hundreds place. The 4 is worth less than the 1 and the 5. The 3 is worth more than the 1. What is the number? _____
7. 0, 2, 3, 6 The 2 is in the tens place. The 3 is in the thousands place. The 6 is worth more than the 2 but less than the 3. What is the number? _____	**8.** 3, 3, 6, 9 It is an odd number. The two 3's are next to each other. The 9 is in the hundreds place. What is the number? _____
9. 1, 1, 7, 8 The number begins and ends with odd numbers. The 8 is in the hundreds place. The two 1's are next to each other. What is the number? _____	**10.** 3, 3, 5, 8 The number begins and ends with odd numbers. The 3's are not next to each other. Both 3's are worth more than the 5. The 8 is in the hundreds place. What is the number? _____
11. 1, 5, 6, 8 It is an even number. The 6 is worth more than the 8. The 5 is in the tens place. The 1 is worth more than the 6. What is the number? _____	**12.** 1, 1, 1, 6 It is an odd number. The 6 is in the hundreds place. What is the number? _____

Rounding Numbers

Practice 16

Rounding numbers to the nearest tens, hundreds, or thousands is useful when adding several numbers together to reach a total. The total will be an estimate of the actual total.

> Round each number to the nearest tens. If the number in the ones place is 0, 1, 2, 3, or 4, round <u>down</u> to the nearest tens. If the number in the ones place is 5, 6, 7, 8, or 9, round <u>up</u> to the nearest tens.
>
> **Example:** 29 → The number in the ones place is a 9, so round up to 30.
>
> **Example:** 22 → The number in the ones place is a 2, so round down to 20.

Directions: Round each number to the nearest 10.

1. 37 _____
2. 25 _____
3. 41 _____
4. 62 _____
5. 73 _____
6. 91 _____

> Round each number to the nearest hundreds. If the number in the tens place is 0, 1, 2, 3, or 4, round <u>down</u> to the nearest hundreds. If the number in the tens place is 5, 6, 7, 8, or 9, round <u>up</u> to the nearest hundreds.
>
> **Example:** 164 → The number in the tens place is a 6, so round up to 200.
>
> **Example:** 131 → The number in the tens place is a 3, so round down to 100.

Directions: Round each number to the nearest 100.

7. 321 _____
8. 578 _____
9. 491 _____
10. 610 _____
11. 847 _____
12. 902 _____

> Round each number to the nearest thousands. If the number in the hundreds place is 0, 1, 2, 3, or 4, round <u>down</u> to the nearest thousands. If the number in the hundreds place is 5, 6, 7, 8, or 9, round <u>up</u> to the nearest thousands.
>
> **Example:** 1,730 → The number in the hundreds place is 7, so round up to 2,000.
>
> **Example:** 1,198 → The number in the hundreds place is 1, so round down to 1,000.

Directions: Round each number to the nearest 1,000.

13. 8,190 _____
14. 7,652 _____
15. 3,943 _____
16. 5,725 _____
17. 9,209 _____
18. 1,871 _____

Rounding Numbers

Practice 17

When rounding to the nearest ten, check the number in the ones place.
- If the number is 5 or greater, round up. Example: 4<u>7</u> → 50
- If the number is 4 or less, round down. Example: 4<u>3</u> → 40

Directions: Round each number to the nearest ten.

1. 85 _____	7. 6 _____	13. 82 _____	19. 24 _____
2. 19 _____	8. 14 _____	14. 7 _____	20. 17 _____
3. 73 _____	9. 69 _____	15. 13 _____	21. 5 _____
4. 15 _____	10. 46 _____	16. 1 _____	22. 2 _____
5. 9 _____	11. 48 _____	17. 59 _____	23. 11 _____
6. 27 _____	12. 31 _____	18. 3 _____	24. 8 _____

When rounding to the nearest hundred, check the number in the tens place.
- If the number is 5 or greater, round up. Example: 1<u>5</u>8 → 200
- If the number is 4 or less, round down. Example: 1<u>3</u>2 → 100

Directions: Round each number to the nearest hundred.

25. 176 _____	31. 295 _____	37. 417 _____	43. 323 _____
26. 451 _____	32. 945 _____	38. 897 _____	44. 183 _____
27. 108 _____	33. 557 _____	39. 827 _____	45. 442 _____
28. 378 _____	34. 774 _____	40. 526 _____	46. 313 _____
29. 565 _____	35. 163 _____	41. 996 _____	47. 834 _____
30. 169 _____	36. 684 _____	42. 406 _____	48. 385 _____

When rounding to the nearest thousand, check the number in the hundreds place.
- If the number is 5 or greater, round up. Example: 1,<u>9</u>03 → 2,000
- If the number is 4 or less, round down. Example: 1,<u>4</u>99 → 1,000

Directions: Round each number to the nearest thousand.

49. 5,929 _____	55. 1,149 _____	61. 7,028 _____	67. 4,359 _____
50. 7,463 _____	56. 1,771 _____	62. 8,932 _____	68. 2,449 _____
51. 1,043 _____	57. 6,416 _____	63. 8,422 _____	69. 2,822 _____
52. 2,957 _____	58. 5,580 _____	64. 3,155 _____	70. 7,527 _____
53. 1,781 _____	59. 1,436 _____	65. 8,993 _____	71. 2,693 _____
54. 6,645 _____	60. 9,297 _____	66. 7,210 _____	72. 3,004 _____

Rounding More Numbers

Practice 18

Directions: Decide whether to round the number to the nearest tens, hundreds, or thousands. Circle the "Th" (thousand); "H" (hundred); or "T" (ten) to show which place the item was rounded to. Write the rounded number in the box. The first one has already been done for you.

1. 298   Th (H) T   [300]
2. 28   Th H T   [   ]
3. 1,092 Th H T   [   ]
4. 11   Th H T   [   ]
5. 817   Th H T   [   ]

6. 2,374 Th H T   [   ]
7. 52   Th H T   [   ]
8. 4,026 Th H T   [   ]
9. 730   Th H T   [   ]
10. 5,963 Th H T   [   ]

Directions: Round each number to the nearest 100. Add or subtract. The first one has been done for you.

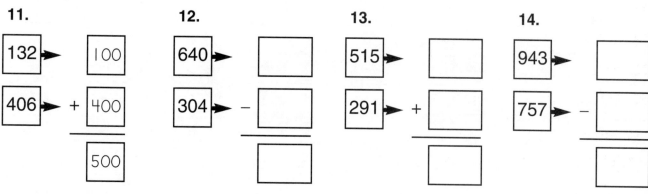

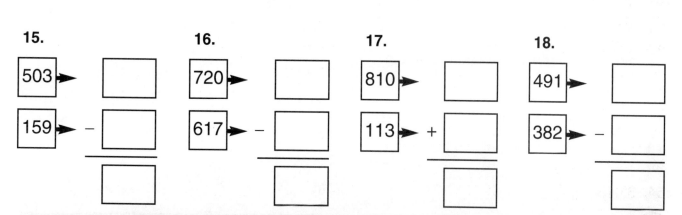

Expanded Notation (Ten Thousands)

Practice 19

Directions: Rewrite the following numbers in expanded notation using only numbers.

Example: 24,478   20,000 + 4,000 + 400 + 70 + 8

1. 51,985 _____
2. 32,156 _____
3. 62,693 _____
4. 92,464 _____
5. 97,034 _____

Directions: Rewrite the following numbers in expanded notation using numbers and words.

Example: 37,114   3 ten thousands + 7 thousands + 1 hundred + 1 ten + 4 ones

6. 47,816 _____
7. 10,360 _____
8. 48,572 _____
9. 92,952 _____
10. 29,108 _____

Directions: Rewrite the following numbers in expanded notation using only words.

Example: 83,580   eight ten thousands + three thousands + five hundreds + eight tens

11. 47,315 _____
12. 10,156 _____
13. 81,693 _____
14. 93,671 _____
15. 12,659 _____

Directions: Rewrite the following numbers in standard form.

Example: fourteen thousand, six hundred fifteen   14,615

16. twenty-four thousand, three hundred eighty-four _____
17. eighty-six thousand, two hundred eighty-one _____
18. sixteen thousand, eight hundred twenty-five _____
19. forty-seven thousand, four hundred sixty-one _____
20. ninety-three thousand, two hundred ninety _____

Place Value (Ten Thousands)

Practice 20

	Ten Thousands	Thousands	Hundreds	Tens	Ones
ten thousand, fifty-nine	1	0,	0	5	9
10,000 + 2,000 + 900 + 40 + 9	1	2,	9	4	9
five ten thousands	5	0,	0	0	0

Directions: Rewrite the numbers that are written incorrectly. Leave the line blank if it is correct.
Example: 7,8172 <u>78,172</u>

1. 310,66 _____
2. 17,523 _____
3. 58,21 _____
4. 4,06 _____
5. 96,249 _____
6. 44,803 _____
7. 5,6 _____
8. 1079,8 _____
9. 1,635 _____
10. 435 _____
11. 29,165 _____
12. 1,085 _____
13. 83,67 _____
14. 12,975 _____
15. 344 _____

Directions: Circle the largest digit. Write its place value.
Example: 26,4⑦1 <u>tens</u>

16. 92,565 _____
17. 10,811 _____
18. 53,383 _____
19. 19,373 _____
20. 25,263 _____
21. 48,074 _____
22. 17,632 _____
23. 11,597 _____
24. 90,158 _____
25. 22,563 _____

Directions: Write the place value of the circled number.
Example: 35,79⑨ <u>ones</u>

26. ④6,420 _____
27. 5①,045 _____
28. ⑦4,221 _____
29. 8⑤,692 _____
30. 91,⑥90 _____
31. ④1,236 _____
32. ③1,679 _____
33. 10,④37 _____
34. ④3,795 _____
35. 1③,361 _____

More Practice with Less and More

Practice 21

Directions: Write the numbers that are ten less and ten more.

1. _____ 21,785 _____
2. _____ 46,887 _____
3. _____ 90,124 _____
4. _____ 62,946 _____
5. _____ 60,616 _____
6. _____ 14,748 _____
7. _____ 15,192 _____
8. _____ 37,589 _____
9. _____ 72,089 _____
10. _____ 16,968 _____
11. _____ 28,542 _____
12. _____ 35,771 _____

Directions: Write the numbers that are one hundred less and one hundred more.

13. _____ 80,486 _____
14. _____ 95,470 _____
15. _____ 92,374 _____
16. _____ 22,561 _____
17. _____ 42,390 _____
18. _____ 46,604 _____
19. _____ 51,677 _____
20. _____ 98,110 _____
21. _____ 86,664 _____
22. _____ 13,249 _____
23. _____ 38,788 _____
24. _____ 33,522 _____

Directions: Write the numbers that are one thousand less and one thousand more.

25. _____ 71,803 _____
26. _____ 70,810 _____
27. _____ 10,252 _____
28. _____ 23,641 _____
29. _____ 35,956 _____
30. _____ 54,684 _____
31. _____ 57,372 _____
32. _____ 46,119 _____
33. _____ 45,293 _____
34. _____ 53,862 _____
35. _____ 79,046 _____
36. _____ 78,979 _____

Adding Five Digits Without Regrouping

Practice 22

Directions: Solve each problem.

1. 42,920
   + 27,025

2. 19,281
   + 60,715

3. 74,337
   + 20,460

4. 70,315
   + 27,574

5. 33,186
   + 62,813

6. 18,897
   + 61,102

7. 85,640
   + 13,249

8. 51,548
   + 27,001

9. 76,228
   + 23,750

10. 89,198
    + 10,301

11. 35,145
    + 41,741

12. 39,372
    + 30,417

13. 28,630
    + 41,336

14. 16,610
    + 40,029

15. 40,455
    + 59,514

16. 45,826
    + 41,173

17. 38,855
    + 50,142

18. 24,608
    + 71,391

19. 58,265
    + 31,724

20. 43,572
    + 41,027

Directions: Solve each word problem

21. If 45,385 people live in Burgundyville and 23,504 people live in Maroontown, what is the combined population of both towns? _____ is the combined population.	22. If 43,723 people attended the fair on Friday and 21,273 people attended the fair on Saturday, what was the total attendance at the fair? _____ was the total attendance.
23. If 70,183 tickets were sold on line and 26,715 tickets were sold at the counter, how many tickets were sold in all? _____ tickets were sold in all.	24. If 16,168 people participated in the bike race and 42,731 people participated in the walk-a-thon, how many people in all participated in the events? _____ people in all.

Subtracting Five Digits Without Regrouping

Practice 23

Directions: Solve each problem.

1. 76,555 − 70,014
2. 75,850 − 64,810
3. 34,500 − 21,000
4. 97,953 − 86,242

5. 63,578 − 33,108
6. 31,762 − 10,462
7. 91,993 − 21,652
8. 53,049 − 53,019

9. 55,037 − 24,016
10. 68,168 − 12,043
11. 68,261 − 44,061
12. 33,228 − 22,018

13. 78,724 − 20,513
14. 37,298 − 26,196
15. 69,644 − 18,203
16. 77,032 − 17,011

17. 68,773 − 68,001
18. 54,408 − 53,208
19. 68,263 − 57,261
20. 77,595 − 16,083

Directions: Solve each word problem.

21. The House of Nuts had 77,989 walnuts. If 26,712 walnuts were packed and shipped, how many walnuts were left? _____ walnuts were left.	22. If 54,323 almonds were gathered and 43,322 of the almonds were roasted, how many almonds were not roasted? _____ almonds were not roasted.
23. There were 45,883 peanuts picked and 42,882 of the peanuts were made into peanut butter. How many peanuts were left? _____ peanuts were left.	24. If 33,152 acorns were gathered and 32,150 of the acorns were planted to become trees, how many acorns were not planted? _____ acorns were not planted.

Adding and Subtracting Five Digits Without Regrouping

Practice 24

Directions: Solve each problem.

1. 99,250
   − 13,050

2. 17,640
   + 62,145

3. 63,830
   − 42,710

4. 11,025
   + 31,464

5. 72,377
   + 25,311

6. 28,624
   − 17,503

7. 89,328
   − 52,104

8. 11,478
   + 88,221

9. 16,354
   + 50,533

10. 53,819
    + 23,020

11. 41,567
    − 41,224

12. 90,961
    − 70,850

13. 12,544
    + 76,324

14. 16,769
    + 72,210

15. 68,747
    − 38,424

16. 70,184
    − 60,073

17. 44,853
    − 23,633

18. 49,336
    + 30,421

19. 48,626
    − 47,313

20. 25,253
    − 14,153

21. 62,896
    − 42,796

22. 87,664
    − 47,612

23. 46,850
    + 10,129

24. 20,246
    + 61,742

25. 10,929
    + 88,050

26. 70,461
    + 19,117

27. 69,290
    − 48,180

28. 74,853
    − 34,621

29. 43,431
    − 32,420

30. 19,233
    + 50,316

31. 93,765
    − 63,710

32. 30,427
    + 33,362

33. 37,354
    − 22,241

34. 92,400
    − 51,300

35. 60,804
    + 31,192

36. 11,604
    + 14,321

Adding Five Digits with Regrouping

Practice 25

Directions: Solve each problem.

1.	32,123 + 15,518	2.	21,738 + 33,495	3.	33,129 + 16,672	4.	36,880 + 11,541
5.	17,036 + 22,155	6.	34,388 + 15,964	7.	58,755 + 18,993	8.	39,229 + 28,858
9.	12,109 + 41,612	10.	10,634 + 41,587	11.	26,455 + 46,725	12.	21,919 + 31,013
13.	41,073 + 13,764	14.	47,218 + 25,133	15.	59,680 + 14,493	16.	24,011 + 27,149
17.	12,563 + 55,553	18.	27,017 + 33,381	19.	53,910 + 18,156	20.	27,697 + 11,194

Directions: Write the numbers in order from smallest to largest.

21.	51,064	49,898	37,123	36,465
22.	72,315	18,012	49,583	23,427
23.	97,381	40,814	57,761	23,323
24.	83,319	10,989	88,901	76,624
25.	66,447	64,219	65,110	27,457
26.	88,110	24,265	72,948	31,032

Column Addition

Practice 26

Directions: Add each set of numbers.

1.	189 12,742 + 9,159	2.	7,216 49,581 + 22	3.	29,578 2,370 + 1,614	4.	6,348 80,163 + 10
5.	57,101 869 + 5,569	6.	36,703 34,140 + 738	7.	41,478 2,241 + 1,229	8.	12,102 18,469 + 5,699
9.	74,556 4,673 + 13,333	10.	510 9,833 + 12,418	11.	766 9,833 + 12,418	12.	766 65,531 + 8,572
13.	8,153 51,896 + 1,077	14.	338 6,935 + 29,310	15.	7,158 3,121 + 19,679	16.	4,201 2,743 + 3,827

Directions: Solve each word problem.

17. Bridget counted 10,771 red toothpicks, 9,423 blue toothpicks, and 18,631 yellow toothpicks. How many toothpicks in all did Bridget count?

    Bridget counted _____ toothpicks in all.

18. Carson sold 46,129 white toothpicks, 9,211 pink toothpicks, and 10,743 brown toothpicks. How many toothpicks did Carson sell in all?

    Carson sold _____ toothpicks in all.

19. Mariah had 6,782 green toothpicks, 849 red toothpicks, and 4,206 yellow toothpicks. How many toothpicks does Mariah have in all?

    Mariah has _____ toothpicks in all.

20. Challenge: Gage had 26,101 toothpicks—4,998 were green, 435 were purple, and the rest were black. How many toothpicks were black?

    There were _____ black toothpicks.

Subtracting Five Digits with Regrouping and Comparing Numbers

Practice 27

Directions: Solve each problem.

1. 65,833 − 54,875
2. 84,915 − 17,349
3. 76,258 − 61,172
4. 54,722 − 39,726

5. 68,195 − 11,390
6. 42,389 − 28,761
7. 36,142 − 16,944
8. 77,810 − 53,109

9. 93,134 − 26,218
10. 44,536 − 10,956
11. 63,516 − 44,369
12. 57,470 − 36,889

13. 29,117 − 10,377
14. 78,748 − 59,255
15. 89,221 − 28,711
16. 67,103 − 52,910

17. 66,238 − 57,955
18. 81,036 − 56,378
19. 91,918 − 69,304
20. 79,422 − 10,142

Directions: Use the symbols < (less than), > (greater than), or = (equal to) to compare each pair of numbers.

21. 10,622 ◯ 87,578
22. 48,445 ◯ 10,710
23. 76,148 ◯ 51,281
24. 53,910 ◯ 62,696
25. 52,322 ◯ 91,175
26. 94,415 ◯ 10,738

27. 12,217 ◯ 15,709
28. 77,332 ◯ 79,942
29. 16,559 ◯ 66,491
30. 94,833 ◯ 81,813
31. 38,902 ◯ 94,942
32. 19,141 ◯ 36,856

Practice 28

Directions: Read the number in the written form. Then write the number in the expanded form and also in standard form.

Written Form	Expanded Form	Standard Form
1. three hundred forty-two thousand, two hundred ten	300,000 + 40,000 + 2,000 + 200 + 10	342,210
2. three hundred forty-five thousand, sixteen		
3. five hundred thirty thousand, two hundred one		
4. seven hundred fifty thousand, nine hundred eleven		
5. four hundred seventy-six thousand, eight hundred twenty		
6. one hundred thousand, four hundred thirty-seven		
7. eight hundred sixty-one thousand, one hundred ninety-two		

Using Expanded Form with Words

Practice 29

> Numbers can also be written in expanded form using numbers and words.
> **Examples**
> 589 is the same as 5 hundreds + 8 tens + 9 ones
> 85,901 is the same as 8 ten thousands + 5 thousands + 9 hundreds + 1 one

Directions: Write the following numbers in expanded form using numbers and words.

1. 623 _____

2. 5,012 _____

3. 30,968 _____

4. 208 _____

5. 73,997 _____

6. 8,647 _____

7. 356,911 _____

8. 415,827 _____

9. 442 _____

10. 6,928 _____

Directions: Write the numbers in order from smallest to greatest.

_____ , _____ , _____ , _____ , _____ ,
_____ , _____ , _____ , _____ ,

32 #8603 Practice Makes Perfect: Place Value   © Teacher Created Materials, Inc.

Practice 30

Decimals are used in writing fractions and in showing different amounts of money.

Example

$\frac{2}{10}$ is the same as .2; both are read as two tenths.

25¢ is the same as $0.25; both are read as twenty-five cents.

The numbers to the left of the decimal point show whole numbers.

The numbers to the right of the decimal point show a fraction of a number.

tens	ones	decimal point	tenths	hundredths
1	4	.	3	5

This number has
1 ten (10.0)
4 ones (4.0)
3 tenths (.3)
5 hundredths (.05)

Directions: Look at each number. Circle the correct way to read the number. The first one has already been done for you.

1. .8   eight hundredths
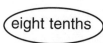
(eight tenths)

2. .01   one hundredth

   one tenth

3. .9   nine tenths

   nine hundredths

4. .06   six tenths

   six hundredths

5. .75   seventy-five hundredths

   seventy-five tenths

6. .42   forty-two tenths

   forty-two hundredths

Directions: Write the number using numerals.

7. 3 tenths _____

8. 29 hundredths _____

9. 7 hundredths _____

10. 81 hundredths _____

11. 4 tenths _____

12. 5 hundredths _____

13. 6 hundredths _____

14. 1 tenth _____

Using a Decimal Point

Practice 31

The **decimal point** is used to separate whole amounts from partial amounts. When reading a number word, the word *and* replaces the decimal point.

	Thousands	Hundreds	Tens	Ones	Decimal	Tenths	Hundredths
Five **and** two-tenths				5	.	2	
Three-hundredths					.	0	3
Ten **and** twelve-hundredths			1	0	.	1	2

Directions: Write the following amounts using a decimal point.

1. three-tenths _____
2. eight-tenths _____
3. nine-tenths _____
4. one-tenth _____
5. four-tenths _____
6. seven-tenths _____
7. two-tenths _____
8. five-tenths _____
9. eight-hundredths _____
10. six-hundredths _____
11. four-hundredths _____
12. three-hundredths _____
13. nine-hundredths _____
14. seven-hundredths _____
15. nine and eighteen-hundredths _____
16. six and nine-hundredths _____
17. fifty-one hundredths _____
18. eight and two-hundredths _____
19. five and eight-tenths _____
20. seven and seventy-nine hundredths _____
21. two and four-hundredths _____
22. five and eight-hundredths _____
23. ten and seventy-one hundredths _____
24. six and forty-three hundredths _____
25. four and seven-hundredths _____
26. three and thirty-four hundredths _____
27. six and three-hundredths _____
28. nine and nine-hundredths _____

Directions: Rewrite the following amounts using only words.

29. .54 _____
30. 9.66 _____
31. 2.18 _____
32. .13 _____
33. .04 _____
34. 10.5 _____
35. .72 _____
36. 5.07 _____

Practice 32

Directions: Use the symbols > (greater than), < (less than), or = (equal to) to compare the two decimal amounts.

1. .54 ◯ .51	2. .33 ◯ .09	3. .91 ◯ .10
4. .71 ◯ .67	5. .61 ◯ .30	6. .75 ◯ .01
7. .91 ◯ .75	8. .15 ◯ .44	9. .08 ◯ .92
10. .59 ◯ .10	11. .80 ◯ .48	12. .22 ◯ .10
13. .98 ◯ .08	14. .30 ◯ .36	15. .66 ◯ .53
16. .73 ◯ .92	17. .06 ◯ .40	18. .25 ◯ .46

Directions: Add or subtract.

19. .42 + .48

20. .07 + .96

21. .29 − .19

22. .15 + .83

23. .79 + .80

24. .26 + .47

25. 7.34 − 3.33

26. 10.48 − 3.92

27. 1.46 + 9.17

28. 11.60 + 21.60

29. 5.34 + 6.10

30. 1.50 + 5.88

31. 2.72 − .65

32. .82 + 6.93

33. .75 − .54

34. 4.10 − 1.71

35. .57 − .41

36. .16 + 4.98

37. 3.91 − 2.04

38. 2.04 + 97.81

39. 98.19 − 43.38

40. 43.38 − 7.26

41. 7.06 − 6.35

42. 7.53 − 2.83

Adding and Subtracting Decimals (Money)

Practice 33

Directions: Solve each problem.

1. A
    $31.02
    + 0.88

2. B
    $ 5.91
    + 61.26

3. C
    $27.30
    − 5.46

4. D
    $97.64
    − 84.95

5. E
    $1.21
    + 4.99

6. F
    $ 6.63
    + 10.53

7. G
    $67.48
    − 5.91

8. H
    $7.21
    + 8.84

9. I
    $0.31
    + 8.51

10. J
    $10.72
    − 9.10

11. K
    $4.55
    + 1.32

12. L
    $4.38
    + 7.79

13. M
    $23.10
    − 22.62

14. N
    $76.46
    − 0.92

15. O
    $0.99
    + 6.89

16. R
    $6.35
    − 0.91

17. S
    $33.57
    − 2.97

18. T
    $5.17
    + 5.32

19. W
    $42.43
    − 8.51

20. Y
    $8.16
    + 7.76

Directions: Write the letter that goes with each amount on the line.

___ ___ ___ ___ ___
$0.48 $7.88 $75.54 $6.20 $15.92

                              ,

___ ___ ___ ___ ___ ___
$12.69 $7.88 $6.20 $30.60 $75.54 $10.49

___ ___ ___ ___ ___ ___
$61.57 $5.44 $7.88 $33.92 $7.88 $75.54

___ ___ ___ ___ ___
$10.49 $5.44 $6.20 $6.20 $30.60

Practice 34

Money Riddles

Directions: Find the total amount of money in each row.

	Quarters	Dimes	Nickels	Pennies	Total Amount
Example	5	4	1	3	$1.73
1.	10	9	2	1	
2.	4	2	7	3	
3.	5	4	3	10	
4.	9	7	8	4	
5.	8	3	4	5	
6.	3	10	1	6	
7.	1	5	9	7	
8.	7	8	6	2	
9.	2	6	10	8	
10.	6	1	5	9	

Directions: Solve each riddle. Use the > (greater than), < (less than), or = (equal to) symbols to compare the amounts.

11. Jane has 5 quarters, 4 dimes, 2 nickels, and 4 pennies. Jack has 1 quarter, 4 dimes, 2 nickels, and 2 pennies. Who has more money?

    Jane ◯ Jack

12. David has 2 quarters, 6 dimes, 2 nickels, and 6 pennies. Jaime has 8 quarters, 1 dime, 5 nickels, and 5 pennies. Who has more money?

    David ◯ Jaime

13. Terri has 2 quarters, 5 dimes, 10 nickels, and 9 pennies. Marlon has 7 quarters, 1 dime, 2 nickels, and 1 penny. Who has more money?

    Terri ◯ Marlon

14. Jade has 4 quarters, 3 dimes, 4 nickels, and 5 pennies. Wade has 2 quarters, 1 dime, 2 nickels, and 1 penny. Who has more money?

    Jade ◯ Wade

15. Taylor has 9 quarters, 10 dimes, 3 nickels, and 7 pennies. Sydney has 4 quarters, 3 dimes, 9 nickels, and 3 pennies. Who has more money?

    Taylor ◯ Sydney

16. Carlos has 1 quarter, 1 dime, 10 nickels, and 6 pennies. Angel has 7 quarters, 6 dimes, 6 nickels, and 7 pennies. Who has more money?

    Carlos ◯ Angel

Working with Integers

Practice 35

An **integer** is a positive or negative whole number. The numbers to the left of the zero are negative numbers. The numbers to the right of the zero are positive numbers.

Directions: Find each number on the number line in relation to zero. Is it to the *left* or to the *right* of the zero? Write *left* or *right* on the line.

Left ← -10 -9 -8 -7 -6 -5 -4 -3 -2 -1 0 1 2 3 4 5 6 7 8 9 10 → Right

1. 2 _____
2. -6 _____
3. 1 _____
4. -9 _____
5. 7 _____
6. -3 _____
7. -1 _____

8. -5 _____
9. 10 _____
10. 3 _____
11. 5 _____
12. -2 _____
13. 8 _____
14. 9 _____

15. -7 _____
16. -4 _____
17. 6 _____
18. -8 _____
19. 4 _____
20. -10 _____

Directions: Use the number line to find each number.

Left ← -10 -9 -8 -7 -6 -5 -4 -3 -2 -1 0 1 2 3 4 5 6 7 8 9 10 → Right

21. Start at 0. Go left 3.	22. Start at 9. Go left 9.	23. Start at 7. Go left 1.
24. Start at -5. Go right 5.	25. Start at -8. Go right 4.	26. Start at -2. Go right 4.
27. Start at -4. Go right 10. Go left 4.	28. Start at 4. Go left 5. Go right 8.	29. Start at -10. Go right 9. Go left 7.
30. Start at -3. Go right 10. Go left 2.	31. Start at -1. Go left 3. Go right 8.	32. Start at 6. Go left 9. Go left 6.

Practice 36

Directions: Use the number lines to solve each problem.

Example	1.
-3 + 5 = 2	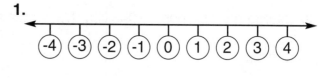 -3 + 2 = _____
2. 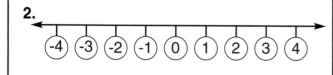 -3 + 3 = _____	3. 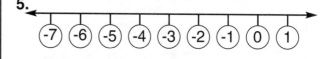 -2 + 2 = _____
4. 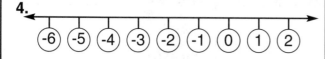 -5 + 2 = _____	5. 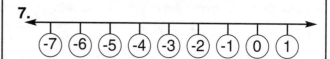 -6 + 4 = _____
6. 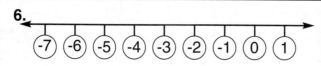 -6 + 6 = _____	7. 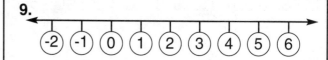 -4 + 0 = _____
8. 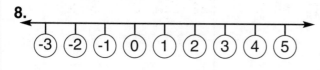 -2 + 3 = _____	9. 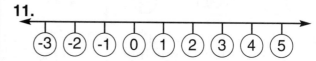 -1 + 5 = _____
10. 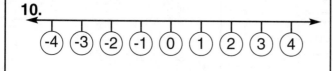 -4 + 3 = _____	11. 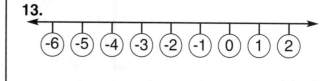 -1 + 1 = _____
12. 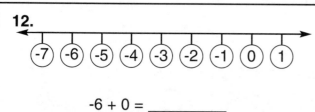 -6 + 0 = _____	13. -5 + 1 = _____

Test Practice 1

Directions: Fill in the circle under the correct answer.

1. Name the number.

   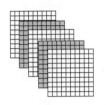

   504   514   524
   Ⓐ   Ⓑ   Ⓒ

2. Name the number.

   349   397   319
   Ⓐ   Ⓑ   Ⓒ

3. Name the place value of the underlined digit.

   8,3<u>2</u>5

   tens   hundreds   thousands
   Ⓐ   Ⓑ   Ⓒ

4. Name the place value of the underlined digit.

   <u>7</u>,641

   tens   hundreds   thousands
   Ⓐ   Ⓑ   Ⓒ

5. Name the place value of the underlined digit.

   3,<u>5</u>27

   tens   hundreds   thousands
   Ⓐ   Ⓑ   Ⓒ

6. Name the digit in the tens place.

   8,796

   7   9   8
   Ⓐ   Ⓑ   Ⓒ

7. Name the digit in the thousands place.

   7,325

   3   7   8
   Ⓐ   Ⓑ   Ⓒ

8. Name the digit in the hundreds place.

   7,648

   1   4   6
   Ⓐ   Ⓑ   Ⓒ

9. What is the largest number that can be made using the digits 4, 9, 6, 5?

   9,654   4,569   9,564
   Ⓐ   Ⓑ   Ⓒ

10. What is the smallest number that can be made using the digits 7, 2, 1, 8?

    7,218   1,278   2,178
    Ⓐ   Ⓑ   Ⓒ

11. Write the number in standard form.

    4 thousands + 3 hundreds + 8 tens + 6 ones

    4,30,86   43,86   4,386
    Ⓐ   Ⓑ   Ⓒ

12. Write the number in standard form.

    2,000 + 100 + 20 + 7

    2,127   2,217   7,212
    Ⓐ   Ⓑ   Ⓒ

Test Practice 2

Directions: Fill in the circle under the correct answer.

---

1. Name the number that is ten more.

752

- 742 (A)
- 762 (B)
- 852 (C)

2. Name the number that is ten less.

630

- 620 (A)
- 640 (B)
- 625 (C)

3. Name the number that is one hundred less.

8,002

- 7,992 (A)
- 7,902 (B)
- 8,102 (C)

---

4. Name the number that is one hundred more.

2,621

- 1,621 (A)
- 2,521 (B)
- 2,721 (C)

5. Name the number.

- 3,281 (A)
- 3,821 (B)
- 2,281 (C)

6. Name the number.

- 1,423 (A)
- 1,243 (B)
- 1,322 (C)

---

7. Compare the numbers.

6,091 ◯ 6,104

- > (A)
- < (B)
- = (C)

8. Compare the numbers.

4,718 ◯ 4,479

- > (A)
- < (B)
- = (C)

9. Complete the pattern.

262, 272, 282, _____

- 302 (A)
- 292 (B)
- 252 (C)

---

10. Add.

3,764
+ 2,105

- 5,869 (A)
- 5,661 (B)
- 5,861 (C)

11. Add.

2,051
+ 3,610

- 5,869 (A)
- 5,661 (B)
- 5,861 (C)

12. Add.

2,648
+ 4,031

- 2,679 (A)
- 6,677 (B)
- 6,679 (C)

Test Practice 3

Directions: Find the correct answer.

---

1. Add.

$$4,711 \\ +5,198$$

9,999   9,909   9,809
 (A)   (B)   (C)

2. Add.

$$3,925 \\ +2,281$$

6,216   5,206   6,206
 (A)   (B)   (C)

3. Subtract.

$$7,641 \\ -2,759$$

5,882   5,118   4,882
 (A)   (B)   (C)

---

4. Subtract.

$$1,571 \\ -1,243$$

332   328   1,328
 (A)   (B)   (C)

5. Subtract.

$$9,986 \\ -8,244$$

1,742   1,644   1,042
 (A)   (B)   (C)

6. What is the mystery number? Use the numbers 3, 4, 7, and 9.
- The 4 has the least value.
- The 3 has the greatest value.
- The 9 is worth more than the 7 but less than the 3.

3,974   7,943   4,397
 (A)   (B)   (C)

---

7. What is the mystery number? Use the numbers 1, 3, 4, and 6.
- The 1 has the greatest value.
- The 3 has the least value.
- The 4 is worth more than the 6 but less than the 1.

4,631   1,346   1,463
 (A)   (B)   (C)

8. Round the number to the nearest ten.

76

60   80   70
(A)   (B)   (C)

9. Round the number to the nearest hundred.

887

700   900   800
 (A)   (B)   (C)

---

10. Round the number to the nearest thousand.

8,978

8,000   9,000   7,000
 (A)   (B)   (C)

11. Which number is the smallest?

567   536   527

567   536   527
 (A)   (B)   (C)

12. Which number is the largest?

7,725   7,693   7,304

7,725   7,693   7,304
 (A)   (B)   (C)

Test Practice 4

Directions: Find the correct answer.

1. Name the place value of the underlined digit.

    6<u>1</u>,397

    tens hundreds thousands
    Ⓐ   Ⓑ   Ⓒ

2. Name the place value of the underlined digit.

    12,6<u>9</u>3

    tens hundreds thousands
    Ⓐ   Ⓑ   Ⓒ

3. Name the place value of the underlined digit.

    <u>4</u>0,017

    tens thousands ten thousands
    Ⓐ   Ⓑ   Ⓒ

4. Which number is the smallest?

    52,579   52,166   52,774

    52,579   52,166   52,774
    Ⓐ   Ⓑ   Ⓒ

5. Which number is the largest?

    41,222   41,108   41,443

    41,222   41,108   41,443
    Ⓐ   Ⓑ   Ⓒ

6. Which number is the smallest?

    33,185   33,183   33,164

    33,185   33,183   33,164
    Ⓐ   Ⓑ   Ⓒ

7. Which number is written correctly?

    26,910   269,10   2,6910
    Ⓐ   Ⓑ   Ⓒ

8. Which number is written correctly?

    3,0172   301,72   30,172
    Ⓐ   Ⓑ   Ⓒ

9. Write the number in standard form.

    forty-seven thousand, three hundred twenty-eight

    47,3028   47,328   4732,8
    Ⓐ   Ⓑ   Ⓒ

10. Name the number that is ten more.

    32,396

    32,406   32,496   32,386
    Ⓐ   Ⓑ   Ⓒ

11. Name the number that is one hundred less.

    73,429

    83,429   73,529   73,329
    Ⓐ   Ⓑ   Ⓒ

12. Add.

    32,837
    + 26,152
    ———

    58,989   14,725   5,8789
    Ⓐ   Ⓑ   Ⓒ

Test Practice 5

Directions: Find the correct answer.

1. Which one means **seven-tenths**?

   7.0 (A) .7 (B) .07 (C)

2. Which one means **nine-hundredths**?

   9.0 (A) .09 (B) .90 (C)

3. Compare the two decimal amounts.

   .15 ◯ .64

   > (A) < (B) = (C)

4. Compare the two decimal amounts.

   .10 ◯ .02

   > (A) < (B) = (C)

5. Add.

   .75
   + .22

   9.7 (A) .97 (B) .097 (C)

6. Subtract.

   .81
   − .77

   .4 (A) .04 (B) 4.0 (C)

7. Subtract.

   $76.47
   − 6.51

   $69.96 (A) $70.96 (B) $60.96 (C)

8. Add.

   4.49
   + 93.82

   98.31 (A) 97.31 (B) 97.21 (C)

9. How much money?

   1 quarter, 3 dimes, 3 nickels, 8 pennies

   $7.80 (A) $1.78 (B) $0.78 (C)

10. Solve.

    -5 + 4 = _____

    1 (A) -1 (B) 9 (C)

11. Solve.

    -3 + 2 = _____

    1 (A) -1 (B) 19 (C)

12. Solve.

    -3 + 3 = _____

    0 (A) -6 (B) 3 (C)

Answer Sheet

Test Practice 1	Test Practice 2	Test Practice 3
1. Ⓐ Ⓑ Ⓒ	1. Ⓐ Ⓑ Ⓒ	1. Ⓐ Ⓑ Ⓒ
2. Ⓐ Ⓑ Ⓒ	2. Ⓐ Ⓑ Ⓒ	2. Ⓐ Ⓑ Ⓒ
3. Ⓐ Ⓑ Ⓒ	3. Ⓐ Ⓑ Ⓒ	3. Ⓐ Ⓑ Ⓒ
4. Ⓐ Ⓑ Ⓒ	4. Ⓐ Ⓑ Ⓒ	4. Ⓐ Ⓑ Ⓒ
5. Ⓐ Ⓑ Ⓒ	5. Ⓐ Ⓑ Ⓒ	5. Ⓐ Ⓑ Ⓒ
6. Ⓐ Ⓑ Ⓒ	6. Ⓐ Ⓑ Ⓒ	6. Ⓐ Ⓑ Ⓒ
7. Ⓐ Ⓑ Ⓒ	7. Ⓐ Ⓑ Ⓒ	7. Ⓐ Ⓑ Ⓒ
8. Ⓐ Ⓑ Ⓒ	8. Ⓐ Ⓑ Ⓒ	8. Ⓐ Ⓑ Ⓒ
9. Ⓐ Ⓑ Ⓒ	9. Ⓐ Ⓑ Ⓒ	9. Ⓐ Ⓑ Ⓒ
10. Ⓐ Ⓑ Ⓒ	10. Ⓐ Ⓑ Ⓒ	10. Ⓐ Ⓑ Ⓒ
11. Ⓐ Ⓑ Ⓒ	11. Ⓐ Ⓑ Ⓒ	11. Ⓐ Ⓑ Ⓒ
12. Ⓐ Ⓑ Ⓒ	12. Ⓐ Ⓑ Ⓒ	12. Ⓐ Ⓑ Ⓒ

Test Practice 4	Test Practice 5	
1. Ⓐ Ⓑ Ⓒ	1. Ⓐ Ⓑ Ⓒ	
2. Ⓐ Ⓑ Ⓒ	2. Ⓐ Ⓑ Ⓒ	
3. Ⓐ Ⓑ Ⓒ	3. Ⓐ Ⓑ Ⓒ	
4. Ⓐ Ⓑ Ⓒ	4. Ⓐ Ⓑ Ⓒ	
5. Ⓐ Ⓑ Ⓒ	5. Ⓐ Ⓑ Ⓒ	
6. Ⓐ Ⓑ Ⓒ	6. Ⓐ Ⓑ Ⓒ	
7. Ⓐ Ⓑ Ⓒ	7. Ⓐ Ⓑ Ⓒ	
8. Ⓐ Ⓑ Ⓒ	8. Ⓐ Ⓑ Ⓒ	
9. Ⓐ Ⓑ Ⓒ	9. Ⓐ Ⓑ Ⓒ	
10. Ⓐ Ⓑ Ⓒ	10. Ⓐ Ⓑ Ⓒ	
11. Ⓐ Ⓑ Ⓒ	11. Ⓐ Ⓑ Ⓒ	
12. Ⓐ Ⓑ Ⓒ	12. Ⓐ Ⓑ Ⓒ	

Answer Key

Page 4
1. 1 hundred + 8 tens + 2 ones
   100 + 80 + 2
   182
2. 2 hundreds + 5 tens + 1 one
   200 + 50 + 1
   251
3. 4 hundreds + 6 tens + 0 ones
   400 + 60 + 0
   460

Page 5
1. 7 hundreds + 0 tens + 7 ones
   700 + 0 + 7
   707
2. 3 hundreds + 4 tens + 3 ones
   300 + 40 + 3
   343
3. 6 hundreds + 5 tens + 8 ones
   600 + 50 + 8
   658
4. 7 hundreds + 1 ten + 5 ones
   700 + 10 + 5
   715
5. 2 hundreds + 2 tens + 4 ones
   200 + 20 + 4
   224
6. 1 hundred + 7 tens + 6 ones
   100 + 70 + 6
   176

Page 6
1. thousands
2. hundreds
3. tens
4. thousands
5. tens
6. hundreds
7. ones
8. thousands
9. thousands
10. hundreds
11. circle 8, tens
12. circle 4, hundreds
13. circle 7, thousands
14. circle 8, tens
15. circle 8, tens
16. circle 9, thousands
17. circle 7, thousands
18. circle 9, hundreds
19. circle 8, thousands
20. circle 8, tens
21. 5,210—1,025
22. 4,332—2,334
23. 7,630—3,067
24. 6,653—3,566
25. 7,522—2,257
26. 8,644—4,468
27. 5,421—1,245
28. 9,876—6,789

Page 7
1. 5,000 + 500 + 20 + 3
2. 6,000 + 100 + 20 + 4
3. 7,000 + 400 + 30 + 9
4. 6,000 + 200 + 40
5. 8,000 + 200 + 5
6. 3 thousands + 4 hundreds + 1 ten + 1 one
7. 2 thousands + 3 hundreds + 5 tens + 7 ones
8. 7 thousands + 1 hundred + 6 tens + 3 ones
9. 1 thousand + 9 hundreds + 6 tens + 3 ones
10. 4 thousands + 5 hundreds + 3 tens + 4 ones
11. one thousand + five hundreds + seven ones
12. five thousands + nine hundreds + nine ones
13. two thousands + six hundreds + one ten + eight ones
14. seven thousands + eight tens + 7 ones
15. four thousands + nine hundreds + two tens + seven ones
16. 8,519
17. 5,832
18. 1,566
19. 2,693
20. 9,246

Page 8
1. 9,981 (first 9 and second 9 are circled)
2. 588
3. 7,539 (7 and 5 are circled)
4. 438
5. 6,218 (6 and 2 are circled)
6. 1,011 (first 1 and 0 are circled)
7. 5,742 (5 and 7 are circled)
8. 842
9. 2,595 (2 and first 5 are circled)
10. circle 6,168—underline 6
11. circle 4,742—underline 4
12. circle 2,613—underline 2
13. circle 4,393—underline 4
14. circle 1,077—underline 1
15. circle 4,231—underline 4
16. circle 6,751—underline 6
17. circle 7,518—underline 7
18. circle 7,848—underline 7
19. Answers will vary.
20. Answers will vary.

Page 9
1. 784, 786
2. 751, 753
3. 302, 304
4. 325, 327
5. 452, 454
6. 576, 578
7. 217, 219
8. 186, 188
9. 494, 496
10. 942, 944
11. 352, 354
12. 508, 510
13. 220, 222
14. 784, 786
15. 467, 469
16. 878, 880
17. 123, 125
18. 628, 630
19. 863, 883
20. 787, 807
21. 671, 691
22. 539, 559
23. 491, 511
24. 164, 184
25. 819, 839
26. 441, 461
27. 330, 350
28. 794, 814
29. 499, 519
30. 545, 565
31. 115, 135
32. 972, 992
33. 730, 750
34. 627, 647
35. 320, 340
36. 274, 294
37. 5,086 5,106
38. 5,945 5,965
39. 3,784 3,804
40. 9,920 9,940
41. 4,802 4,822
42. 8,152 8,172
43. 1,857 1,877
44. 3,259 3,279
45. 9,532 9,552
46. 5,100 5,120
47. 1,249 1,269
48. 7,399 7,419
49. 8,294 8,314
50. 4,926 4,946
51. 3,756 3,776
52. 1,998 2,018
53. 1,895 1,915
54. 1,252 1,272
55. 5,032 5,232
56. 7,352 7,552
57. 2,903 3,103
58. 3,164 3,364
59. 5,257 5,457
60. 7,118 7,318
61. 1,774 1,974
62. 9,494 9,694
63. 3,253 3,453
64. 4,992 5,192
65. 2,396 2,596
66. 6,013 6,213
67. 7,116 7,316
68. 1,960 2,160
69. 5,646 5,846
70. 6,729 6,929
71. 9,788 9,988
72. 7,719 7,919

Page 10
1. 9 thousands 3 hundreds 0 tens 2 ones
   9,000 + 300 + 0 + 2
   9,302
2. 4 thousands 2 hundreds 3 tens 4 ones
   4,000 + 200 + 30 + 4
   4,234

Page 11
1. 5 thousands 8 hundreds 8 tens 4 ones
   5,000 + 800 + 80 + 4
   5,884
2. 8 thousands 3 tens 7 ones
   8,000 + 30 + 7
   8,037
3. 3 thousands 4 hundreds 6 tens 6 ones
   3,000 + 400 + 60 + 6
   3,466

Page 12
1. <, less
2. >, greater
3. >, greater
4. <, less
5. <, less
6. >, greater
7. <, less
8. <, less
9. <, less
10. <, less
11. >, greater
12. >, greater
13. <, less
14. <, less
15. <, less
16. >, greater
17. <, less
18. 79, 89, 99, 109, 119, 129, 139
19. 42, 52, 62, 72, 82, 92, 102
20. 439, 449, 459, 469, 479, 489, 499
21. 1,005 1,015 1,025 1,035 1,045 1,055 1,065
22. 3,887 3,897 3,907 3,917 3,927 3,937 3,947

Page 13
1. 4,689   11. 9,993
2. 9,985   12. 6,869
3. 8,989   13. 8,999
4. 7,158   14. 7,669
5. 5,988   15. 3,879
6. 6,784   16. 8,999
7. 5,856   17. 5,999
8. 9,891   18. 8,863
9. 4,986   19. 9,759
10. 9,697   20. 8,698
Answers > 5,000:
9,985–8,989–7,158–5,988–6,784–5,856–9,891–9,697–9,993–6,869–8,999–7,669–8,999–5,999–8,863–9,759–8,698
Answers < 5,000:
4,689–4,986–3,879

Page 14
1. 3,050   13. 5,040
2. 8,603   14. 3,005
3. 5,367   15. 4,030
4. 1,442   16. 2,005
5. 2,455   17. 3
6. 2,262   18. 3,910
7. 5,530   19. 1,201
8. 8,015   20. 4,012
9. 1,420   21. 6,000
10. 3,121   22. 2,010
11. 4,003   23. 2,100
12. 2,051   24. 3,153
Group 1:
1,442–2,455–3,050–5,367–8,603

Answer Key

Group 2:
1,420–2,262–3,121–5,530–8,015
Group 3:
2,051–3,005–4,003–4,030–5,040
Group 4:
3–1,201–2,005–3,910–4,012
Group 5:
2,010–2,100–3,153–6,000

Page 15
1. 8,994
2. 1,055
3. 410
4. 1,017
5. 9,359
6. 2,010
7. 120
8. 505
9. 3,512
10. 5,886
11. 5,879
12. 3,066
13. 7,098
14. 4,004
15. 4,175
16. 1,301
17. 3,599
18. 4,569
19. 551
20. 230

21.–25. Math problems will vary. Below are shown some sample problems that can be made.
2,907 + 7,092 = 9,999
9,764 – 9,764 = 0
6,401 + 1,460 = 7,861
3,987 – 3,879 = 108
4,578 + 4,578 = 9,156

Page 16
1. 6,808
2. 7,466
3. 7,615
4. 6,237
5. 10,149
6. 6,353
7. 8,174
8. 9,608
9. 9,558
10. 9,840
11. 9,350
12. 7,076
13. 9,390
14. 9,449
15. 8,819
16. 7,220
17. 7,359
18. 4,782
19. 7,092
20. 7,593

Secret Message:
Numbers are everywhere!

Page 17
1. 1,036
2. 3,020
3. 3,484
4. 262
5. 608
6. 2,209
7. 1,488
8. 1,149
9. 2,505
10. 4,029
11. 575
12. 118
13. 3,179
14. 4,489
15. 4,897
16. 3,606
17. 4,728
18. 993
19. 1,651
20. 7,741
21. 1922 – 1874 = 48
22. 1943 – 1864 = 79
23. 1931 – 1847 = 84
24. 1872 – 1791 = 81
25. 1947 – 1863 = 84
26. 1912 – 1867 = 45

Page 18
1. 1,120
2. 4,995
3. 7,621
4. 2,780
5. 5,374
6. 3,154
7. 3,620
8. 6,933
9. 7,811
10. 3,835
11. 1,658
12. 1,611

Page 19
1. 40
2. 30
3. 40
4. 60
5. 70
6. 90
7. 300
8. 600
9. 500
10. 600
11. 800
12. 900
13. 8,000
14. 8,000
15. 4,000
16. 6,000
17. 9,000
18. 2,000

Page 20
1. 90
2. 20
3. 70
4. 20
5. 10
6. 30
7. 10
8. 10
9. 70
10. 50
11. 50
12. 30
13. 80
14. 10
15. 10
16. 0
17. 60
18. 0
19. 20
20. 20
21. 10
22. 0
23. 10
24. 10
25. 200
26. 500
27. 100
28. 400
29. 600
30. 200
31. 300
32. 900
33. 600
34. 800
35. 200
36. 700
37. 400
38. 900
39. 800
40. 500
41. 1,000
42. 400
43. 300
44. 200
45. 400
46. 300
47. 800
48. 400
49. 6,000
50. 7,000
51. 1,000
52. 3,000
53. 2,000
54. 7,000
55. 1,000
56. 2,000
57. 6,000
58. 6,000
59. 1,000
60. 9,000
61. 7,000
62. 9,000
63. 8,000
64. 3,000
65. 9,000
66. 7,000
67. 4,000
68. 2,000
69. 3,000
70. 8,000
71. 3,000
72. 3,000

Page 21
1. H, 300
2. T, 30
3. Th, 1,000
4. T, 10
5. H, 800
6. Th, 2,000
7. T, 50
8. Th, 4,000
9. H, 700
10. Th, 6,000
11. 100 + 400 = 500
12. 600 – 300 = 300
13. 500 + 300 = 800
14. 900 – 800 = 100
15. 500 – 200 = 300
16. 700 – 600 = 100
17. 800 + 100 = 900
18. 500 – 400 = 100

Page 22
1. 50,000 + 1,000 + 900 + 80 + 5
2. 30,000 + 2,000 + 100 + 50 + 6
3. 60,000 + 2,000 + 600 + 90 + 3
4. 90,000 + 2,000 + 400 + 60 + 4
5. 90,000 + 7,000 + 30 + 4
6. 4 ten thousands + 7 thousands + 8 hundreds + 1 ten + 6 ones
7. 1 ten thousand + 3 hundreds + 6 tens
8. 4 ten thousands + 8 thousands + 5 hundreds + 7 tens + 2 ones
9. 9 ten thousands + 2 thousands + 9 hundreds + 5 tens + 2 ones
10. 2 ten thousands + 9 thousands + 1 hundred + 8 ones
11. four ten thousands + seven thousands + three hundreds + one ten + five ones
12. one ten thousand + one hundred + five tens + six ones
13. eight ten thousands + one thousand + six hundreds + nine tens + three ones
14. nine ten thousands + three thousands + six hundreds + seven tens + one one
15. one ten thousand + two thousands + six hundreds + five tens + nine ones
16. 24,384
17. 86,281
18. 16,825
19. 47,461
20. 93,290

Page 23
1. 31,066
2. ok as is
3. 5,821
4. 406
5. ok as is
6. ok as is
7. 56
8. 10,798
9. ok as is
10. ok as is
11. ok as is
12. ok as is
13. 8,367
14. ok as is
15. ok as is
16. circle 9, ten thousands
17. circle 8, hundreds
18. circle 8, tens
19. circle 9, thousands
20. circle 6, tens
21. circle 8, thousands
22. circle 7, thousands
23. circle 9, tens
24. circle 9, ten thousands
25. circle 6, tens
26. ten thousands
27. thousands
28. ten thousands
29. thousands
30. hundreds
31. ten thousands
32. ten thousands
33. hundreds
34. ten thousands
35. thousands

Page 24
1. 21,775—21,795
2. 46,877—46,897
3. 90,114—90,134
4. 62,936—62,956
5. 60,606—60,626
6. 14,738—14,758
7. 15,182—15,202
8. 37,579—37,599
9. 72,079—72,099
10. 16,958—16,978
11. 28,532—28,552
12. 35,761—35,781
13. 80,386—80,586
14. 95,370—95,570
15. 92,274—92,474
16. 22,461—22,661
17. 42,290—42,490
18. 46,504—46,704
10. 51,577—51,777
20. 98,010—98,210
21. 86,564—86,764
22. 13,149—13,349
23. 38,688—38,888
24. 33,422—33,622
25. 70,803—72,803
26. 69,810—71,810
27. 9,252—11,252
28. 22,641—24,641
29. 34,956—36,956
30. 53,684—55,684
31. 56,372—58,372
23. 45,119—47,119
33. 44,293—46,293
34. 52,862—54,862
35. 78,046—80,046
36. 77,979—79,979

Page 25
1. 69,945
2. 79,996
3. 94,797
4. 97,889
5. 95,999
6. 79,999
7. 98,889
8. 78,549
9. 99,978
10. 99,499
11. 76,886
12. 69,789
13. 69,966
14. 56,639
15. 99,969
16. 86,999
17. 88,997
18. 95,999
19. 89,989
20. 84,599
21. 68,889
22. 64,996
23. 96,898
24. 58,899

Page 26
1. 6,541
2. 11,040
3. 13,500
4. 11,711
5. 30,470
13. 58,211
14. 11,102
15. 51,441
16. 60,021
17. 772

© Teacher Created Materials, Inc.   #8603 Practice Makes Perfect: Place Value

Answer Key

6. 21,300
7. 70,341
8. 30
9. 31,021
10. 56,125
11. 24,200
12. 11,210
18. 1,200
19. 11,002
20. 61,512
21. 51,277
22. 11,001
23. 3,001
24. 1,002

Page 27
1. 86,200
2. 79,785
3. 21,120
4. 42,489
5. 97,688
6. 11,121
7. 37,224
8. 99,699
9. 66,887
10. 76,839
11. 343
12. 20,111
13. 88,868
14. 88,979
15. 30,323
16. 10,111
17. 21,220
18. 79,757
19. 1,313
20. 11,100
21. 20,100
22. 40,052
23. 56,979
24. 81,988
25. 98,979
26. 89,578
27. 21,110
28. 40,232
29. 11,011
30. 69,549
31. 30,055
32. 63,789
33. 15,113
34. 41,100
35. 91,996
36. 25,925

Page 28
1. 47,641
2. 55,233
3. 49,801
4. 48,421
5. 39,191
6. 50,352
7. 77,748
8. 68,087
9. 53,721
10. 52,221
11. 73,180
12. 52,932
13. 54,837
14. 72,351
15. 74,173
16. 51,160
17. 68,116
18. 60,398
19. 72,066
20. 38,891
21. 36,465–37,123–49,898–51,064
22. 18,012–23,427–49,583–72,315
23. 23,323–40,814–57,761–97,381
24. 10,989–76,624–83,319–88,901
25. 27,457–64,219–65,110–66,447
26. 24,265–31,032–72,948–88,110

Page 29
1. 22,090
2. 56,819
3. 33,562
4. 86,521
5. 63,539
6. 71,581
7. 44,948
8. 36,270
9. 92,562
10. 22,761
11. 23,017
12. 74,869
13. 61,126
14. 36,583
15. 29,958
16. 10,771
17. 38,825
18. 66,083
19. 11,837
20. 20,668

Page 30
1. 10,958
2. 67,566
3. 15,086
4. 14,996
5. 56,805
6. 13,628
7. 19,198
8. 24,701
9. 66,916
10. 33,580
11. 19,147
12. 20,581
13. 18,740
14. 19,493
15. 60,510
16. 14,193
17. 8,283
18. 24,658
19. 22,614
20. 69,280
21. <
22. >
23. >
24. <
25. <
26. >
27. <
28. <
29. <
30. >
31. <
32. <

Page 31
1. 300,000 + 40,000 + 2,000 + 200 + 10
   342,210
2. 300,000 + 40,000 + 5,000 + 10 + 6
   345,016
3. 500,000 + 30,000 + 200 + 1
   530,201
4. 700,000 + 50,000 + 900 + 10 + 1
   750,911
5. 400,000 + 70,000 + 6,000 + 800 + 20
   476,820
6. 100,000 + 400 + 30 + 7
   100,437
7. 800,000 + 60,000 + 1,000 + 100 + 90 + 2
   861,192

Page 32
1. 6 hundreds + 2 tens + 3 ones
2. 5 thousands + 1 ten + 2 ones
3. 3 ten thousands + 9 hundreds + 6 tens + 8 ones
4. 2 hundreds + 8 ones
5. 7 ten thousands + 3 thousands + 9 hundreds + 9 tens + 7 ones
6. 8 thousands + 6 hundreds + 4 tens + 7 ones
7. 3 hundred thousands + 5 ten thousands + 6 thousands + 9 hundreds + 1 ten + 1 one
8. 4 hundred thousands + 1 ten thousand + 5 thousands + 8 hundreds + 2 tens + 7 ones
9. 4 hundreds + 4 tens + 2 ones
10. 6 thousands + 9 hundreds + 2 tens + 8 ones
    208–442–623–5,012–6,928–8,647–30,968–73,997–356,911–415,827

Page 33
1. eight tenths
2. one hundredth
3. nine tenths
4. six hundredths
5. seventy-five hundredths
6. forty-two hundredths
7. .3
8. .29
9. .07
10. .81
11. .4
12. .05
13. .06
14. .1

Page 34
1. .3
2. .8
3. .9
4. .1
5. .4
6. .7
7. .2
8. .5
9. .08
10. .06
11. .04
12. .03
13. .09
14. .07
15. 9.18
16. 6.09
17. .51
18. 8.02
19. 5.8
20. 7.79
21. 2.04
22. 5.08
23. 10.71
24. 6.43
25. 4.07
26. 3.34
27. 6.03
28. 9.09
29. fifty-four hundredths
30. nine and sixty-six hundredths
31. two and eighteen-hundredths
32. thirteen-hundredths
33. four-hundredths
34. ten and five-tenths
35. seventy-two hundredths
36. five and seven-hundredths

Page 35
1. >
2. >
3. >
4. >
5. >
6. >
7. >
8. <
9. <
10. >
11. >
12. >
13. >
14. <
15. >
16. <
17. <
18. <
19. .90
20. 1.03
21. .10
22. .98
23. 1.59
24. .73
25. 4.01
26. 6.56
27. 10.63
28. 33.20
29. 11.44
30. 7.38
31. 2.07
32. 7.75
33. .21
34. 2.39
35. .16
36. 5.14
37. 1.87
38. 99.85
39. 54.81
40. 36.12
41. .71
42. 4.70

Page 36
1. $31.90
2. $67.17
3. $21.84
4. $12.69
5. $6.20
6. $17.16
7. $61.57
8. $16.05
9. $8.82
10. $1.62
11. $5.87
12. $12.17
13. $0.48
14. $75.54
15. $7.88
16. $5.44
17. $30.60
18. $10.49
19. $33.92
20. $15.92
Secret message: Money doesn't grow on trees.

Page 37
1. $3.51
2. $1.58
3. $1.90
4. $3.39
5. $2.55
6. $1.86
7. $1.27
8. $2.87
9. $1.68
10. $1.94
11. >
12. <
13. <
14. >
15. >
16. <

Page 38
1. Right
2. Left
3. Right
4. Left
5. Right
6. Left
7. Left
8. Left
9. Right
10. Right
11. Right
12. Left
13. Right
14. Right
15. Left
16. Left
17. Right
18. Left
19. Right
20. Left
21. -3
22. 0
23. 6
24. 0
25. -4
26. 2
27. 2
28. 7
29. -8
30. 5
31. 4
32. -9

Page 39
1. -1
2. 0
3. 0
4. -3
5. -2
6. 0
7. -4
8. 1
9. 4
10. -1
11. 0
12. -6
13. -4

Page 40
1. C
2. A
3. A
4. C
5. B
6. B
7. B
8. C
9. A
10. B
11. C
12. A

Page 41
1. B
2. A
3. B
4. C
5. A
6. B
7. B
8. A
9. B
10. A
11. B
12. C

Page 42
1. B
2. C
3. C
4. B
5. A
6. A
7. C
8. B
9. B
10. B
11. C
12. A

Page 43
1. C
2. A
3. C
4. B
5. C
6. C
7. A
8. C
9. B
10. A
11. C
12. A

Page 44
1. B
2. B
3. B
4. A
5. B
6. B
7. A
8. A
9. C
10. B
11. B
12. A